John Glenn Jr. Debriefed

First Thoughts and Pilot's Debriefing of His Historic *Friendship 7* Flight

SpaceHistory101.com Press

Reprinted with the permission of the journal: "Quest: The History of Spaceflight Quarterly" http:// www.spacehistory101.com

Images courtesy of NASA.

Printed in the United States of America
ISBN: 978-188702298-9

Table of Contents

Acronyms

ALFA	Air Lubricated Free-Attitude Trainer
ASCS	Automatic Stabilization and Control System
CapCom	Capsule Communicator
ECS	Engine Control System
F80	F80 Crusader Aircraft
HF	High Frequency
MCC	Mission Control Center
UHF	Ultra High Frequency
WPA	Works Projects Administration

Preface

John H. Glenn, Jr. became the first American to orbit Earth on February 20, 1962. The following is the preliminary debriefing report conducted aboard the destroyer Noa and at Grand Turk on February 20, 21, and 22. It begins with Glenn's narrative account of the flight based on the recorded debriefing onboard Noa followed by specific questions posed during the debrief. These questions cover a chronological review of the flight from pre-launch entering the capsule all the way through to recovery and relate his experiences during his three-orbit flight to his training along with the design of his pressure suit and capsule.

Glenn relates experiences that are taken for granted today, but were revelations to NASA as a result of his short flight. After the questions, Glenn reported special astronomical, meteorological, and terrestrial observations made during the flight together with predominant sensations during launch and powered flight.

The debriefing team was instructed to not interrupt the astronaut. The astronaut was to read the question and answer on the voice recorder. The team members would take notes and ask all candid questions at the completion of the formal debriefing. This mode of operation was essential in order to obtain a clean record of the debriefing.

This preliminary report was used to aid in the writing of the official NASA post-launch report.

John Herschel Glenn, Jr. Biography

Born: 18 July 1921
Birth City/State: Cambridge, Ohio, United States

Personal

Married, 6 April 1943 to Anna "Annie" Margaret Castor

Children: John David and Carolyn Ann
Father: John Herschel Glenn, Sr
Mother: Teresa Sproat

Education

- Muskingum College, New Concord, Ohio, September 1939-January 1942
- Private Pilot's License, Muskingum College, July 1941
- Naval Aviation Cadet Program at University of Iowa, 1943
- Naval Test Pilot School, Patuxant River, Maryland, July 1954
- University of Maryland, November 1956-April 1959
- B.S. in Mathematics, Muskingum College, 1962

Career Highlights: United States Marine Corps (retired 1 January 1965)

- F4U Pilot, Marine Fighter Squadron 155, Marshall Islands (1943-1944)
- Pilot, Fighter Squadron 218, China and Guam (1946-1948)
- Instructor, Advanced Flight Training School, Corpus Christi, Texas (1949-1951)
- Attended Amphibious Warfare Training, Quantico, Virginia
- Pilot, Marine Fighter Squadron 311, Korea, 1953
- F-86 Sabrejet USAF Exchange Pilot, Korea
- Project Officer, Naval Test Pilot School, Patuxant River, Maryland,
- Fighter Design Branch, Navy Bureau of Aeronautics (1956-1959)
- WWII: 59 combat missions, Korea: 63 combat missions

Career Highlights: NASA

- NASA Space Task Group, Langley Research Center, Hampton, Virginia / NASA Manned Spacecraft Center, Houston, Texas
    - Mercury Astronaut (1959-1964)
    - Payload Specialist STS-95 (January 1998-November 1998)

Career Highlights: Post-NASA

- Executive, Royal Crown International (1965-1974)
- Senator [D-OH], United States Congress, (1974-1999)

Astronaut Career

Mercury Program

Astronaut Group:	1
Mission:	Mercury Atlas 6 / *Friendship 7*
Launch Date:	20 February 1962
Orbits:	3. First American in orbit
Flight Time:	4 hours, 55 minutes, 23 seconds

STS-95

Date:	29 October 1998 – 7 November 1998
Title	Payload Specialist
Orbits:	134
Flight Time:	213 hours, 44 minutes

Awards and Honors Overview

- Distinguished Flying Cross (six)
- Air Medal with 18 Oak Leaf Clusters (for service in World War II and Korea)
- Navy Unit Commendation (for service in Korea)
- Asiatic-Pacific Campaign Medal
- American Campaign Medal
- World War II Victory Medal
- China Service Medal
- National Defense Service Medal
- Korean Service Medal
- United Nations Service Medal
- Korean Presidential Unit Citation
- Navy Astronaut Wings
- Marine Corps Astronaut medal
- NASA Distinguished Service Medal
- Congressional Space Medal of Honor
- Honorary Doctor of Science, Muskingum College, 1961
- Honorary Doctor of Engineering, Nihon University, Tokyo, Japan
- Honorary Doctor of Engineering, Wagner College, New York
- Honorary Doctor of Engineering, New Hampshire College.

Brief Summary of MA-6 Orbital Flight

Based on recorded debriefing onboard the destroyer USS NOA shortly after the MA-6 mission on February 20, 1962.

by John H. Glenn, Jr.

There are many things that are so impressive, it's almost impossible to try and describe the sensations that I had during the flight. I think the thing that stands out more particularly than anything else right at the moment is the reentry during the fireball. I left the shutters open specifically so I could watch it. It got to a brilliant orange color; it was never too blinding. The retropack was still aboard and shortly after reentry began, it started to break up in big chunks. One of the straps came off and came around across the window. There were large flaming pieces of the retropack—I assume that's what they were that broke off and came tumbling around the sides of the capsule. I could see them going on back behind me then and making little smoke trails. I could also see a long trail of what probably was ablation material ending in a small bright spot similar to that in the pictures out of the window taken during the MA-5 flight. I saw the same spot back there and I could see it move back and forth as the capsule oscillated slightly. Yes, I think the reentry was probably the most impressive part of the flight.

Starting back with highlights of the flight: Insertion was normal this morning except for the delays that were occasioned by hatch-bolt trouble and by the microphone fitting breaking off in my helmet. The weather cleared up nicely and after only moderate delays, we got off.

Lift-off was just about as I had expected. There was some vibration. Coming up off the pad, the roll programming was very noticeable, as the spacecraft swung around to the proper azimuth. There also was no doubt about when the pitch programming started. There was some vibration at lift-off from the pad. It smoothed out just moderately; never did get to very smooth flight until we were through the high g area. At this time I would guess a minute and fifteen to twenty seconds it was very noticeable. After this, it really smoothed out and by a minute and a half, or about the time the cabin pressure sealed off, it was as smooth as could be.

The staging was normal, though I had expected a more sharp cutoff. It felt as though the g ramped down for maybe half a second. For some reason, it was not as abrupt as I had anticipated it might be. The accelerometers read one and a quarter g's when I received a confirmation on staging from the Capsule Communicator. I had been waiting for this message at that point because I was set to go to tower jettison as we had planned, in case the booster had not staged. At this time, I also saw a wisp of smoke and I though perhaps the tower had jettisoned early.

The tower really had not jettisoned at that time and did jettison on schedule at 2+34. As the booster and capsule pitched over and the tower jettisoned, I had a first glimpse of the horizon; it was a beautiful sight, looking eastward across the Atlantic.

Toward the last part of the insertion, the vibration began building up again. This I hadn't quite expected; it wasn't too rough but it was noticeable. Cutoff was very good; the capsule acted just as it was suppose to. The ASCS damped and turned the spacecraft around. As we were completing the turnaround, I glanced out of the window and the booster was right there in front of me. It looked as though it wasn't more than a hundred yards away. The small end of the booster was pointing toward the northeast and I saw it a number of times from then on, for about the next seven or eight minutes as it slowly went below my altitude and moved farther away. That was very impressive.

I think I was really surprised at the ease with which the controls check went. It was almost just like making the controls check on the Procedures Trainer that we've done so many times. The control check went off like clockwork; there was no problem at all. Everything damped when it should damp and control was easy. Zero-g was noticeable at SECO. I had a very slight sensation of tumbling forward head-over-heels. It was very slight; not as pronounced an effect as we experience on the centrifuge. During turnaround, I had no sensation of angular acceleration. I acclimated to weightlessness in just a matter of seconds; it was very surprising. I was reaching for switches and doing things and having no problem. I didn't at any time notice any tendency to overshoot a switch. It seemed it's just natural to acclimate to this new condition. It was very comfortable. Under the weightless condition, the head seemed to be a little farther out of the couch which made it a little easier to see the window, though I could not get up quite as near to the window as I thought I might.

The rest of the first orbit went pretty much as planned, with reports to the stations coming up on schedule. I was a little behind at a couple of points but most of the things were going right according to schedule, including remaining on the automatic control system for optimum radar and communications tracking. Sunset from this altitude is tremendous. I had never seen anything like this and it was a truly beautiful, beautiful sight. The speed at which the sun goes down is very remarkable, of course. The brilliant orange and blue layers spread out probably 450 to 600 each side of the sun tapering very slowly toward the horizon. I could not pick up any appreciable Zodiacal light. I looked for it closely; I think perhaps I was not enough right adapted to see it. Sunrise, I picked up in the periscope. At every sunrise, I saw little, specks, brilliant specks, floating around outside the capsule. I have no idea what they were. On the third orbit, I turned around at sunrise so that I could face into the Sun and see if, they were still heading in the same direction and they were. But I noticed them every sunrise and tried to get pictures of them.

Just as I came over Mexico at the end of the first orbit, I had my first indication of the ASCS problem that was to stick with me for the rest of the flight. It started out with the yaw rate going off at about one and one-half degrees per second to the right. The capsule would not stay in orbit mode, but would go out of limits. When it reached about 200 instead of the 300 I expected, it would kick back into orientation mode and swing back with the rate going over into the left yaw to correct back to its normal orbit attitude. Sometimes, it would cross-couple into pitch and roll and we'd go through a general disruption of orbit mode until it settled down into orbit attitude. Then yaw would again start a slow drift to the right and the ASCS would kick out again into orientation mode. I took over manually at that point and from then on, through the rest of the flight, this was my main concern. I tried to pick up the flight plan again at a few points and I accomplished a few more things on it, but I'm afraid most of the flight time beyond that point was taken up with checking the various modes of the ASCS. I did have full control in fly-by-wire and later on during the flight, the yaw problem switched from left to right. It acted exactly the same, except it would drift off to the left instead of the right. It appeared also that any time I was on manual control and would be drifting away from regular orbit attitude for any appreciable period of time that the attitude indications would then be off when I came back to orbit attitude. I called out some of these and I remember that at one time, roll was off 30^0, yaw was off 35^0; and pitch was off 76^0, These were considerable errors and I have no explanation for them at this time. I could control on fly-by-wire and manual very adequately. It was not difficult at all. Fly-by-wire was by far the most accurate means of control, even though I didn't have accurate control in yaw at all times.

Retrorockets were fired right on schedule just off California and it was surprising coming out of the zero-g field that the retrorockets firing felt as though I were accelerating in the other direction back toward Hawaii. However, after retrofire was completed when I could glance out the window again, it was easy to tell, of course, which way I was going, even though my sensations during retrofire had been that I was going in the other direction. I made retrofire on automatic control. Apparently, the solid-on period for slaving just prior to retrofire brought the gyros back up to orbit attitude, because they corrected very nicely during that period. The spacecraft was just about in orbit attitude as I could see it from the window and through the periscope just prior to retrofire. So I feel that we were right in attitude. I left it on ASCS and backed up manually and worked right along with the ASCS during retrofire. I think the retroattitude held almost exactly on and I would guess that we were never more than 30 off in any axis at any time during retrofire

Following retrofire, a decision was made to have me reenter with the retropackage still on because of the uncertainty as to whether the landing bag had been extended. I don't know all the reasons yet for that particular decision, but I assume that it had been pretty well thought out and it obviously was. I punched up .05g manually at a little after the time it was given to me. I was actually in a small g-field at the time I pushed up .05g and it went green and I began to get noise, or what sounded like small things brushing against the capsule. I began to get this very shortly after .05g and this noise, kept increasing. Well before we got into the real heavy fireball area, one strap swung around and hung down over the window. There was some smoke. I don't know whether the bolt fired at the center of the pack or what happened. The capsule kept on its course. I didn't get too far off of reentry attitude. I went to manual control for reentry after the retros fired and had no trouble controlling reentry attitude through the high-g area. Communications blackout started a little bit before the fireball. The fireball was very intense. I left the shutters open the whole time and observed it and it got to be a very, very bright orange color. There were large flaming pieces of what I assume was the retropackage breaking off and going back behind the capsule. This was of some concern, because I wasn't sure of what it was. I had visions of them possibly being chunks of heat shield breaking off, but it turned out it was not that.

The oscillations that built up after peak-g were more than I could control with the manual system. I was damping okay and it just plain overpowered me and I could not do anymore about it. I switched to Aux. Damp as soon as I could raise my arm up after the g-pulse to help damp and this did help some. However, even on Aux. Damp, the capsule was swinging back and forth very rapidly and the oscillations were divergent as we descended to about 35,000 feet. At this point, I elected to try to put the drogue out manually, even though it was high, because I was afraid we were going to get over to such an attitude that the capsule might actually be going small end down during part of the flight if the oscillations kept going the way they were. And just as I was reaching up to pull out the drogue on manual, it came out by itself. The drogue did straighten the capsule out in good shape. I believe the altitude was somewhere between 30,000 and 35,000 at that point.

I came on down; the snorkels, I believe, came out at about 16,000 or 17,000. The periscope came out. There was so much smoke and dirt on the windshield that it was somewhat difficult to see. Every time I came around to the Sun for I had established my roll rate on manual it was virtually impossible to see anything out through the window.

The capsule was very stable when the antenna section jettisoned. I could see the whole recovery system just lined up in one big line as it came out. It unreeled and blossomed normally; all the panels and visors looked good. I was going through my landing checkoff list when the Capsule Communicator called to remind me to deploy the landing bag. I flipped the switch to auto immediately and the green light came on and I felt the bag release. I was able to watch the water coming towards me in the periscope. I was able to estimate very closely when I would hit the water. The impact bag was a heavier shock than I had expected, but it did not bother me.

Communications with the recovery ship *Noa* were very good. The *Noa* had me in sight before impact and estimated 20 minutes to recovery, which turned out to be about right. When the destroyer came alongside, they hooked on with the shepherd's hook and cut the HF antenna. During capsule pickup, I received one good solid bump on the side of the ship as it rolled. Once on deck I took the left hand panel loose and started to disconnect the suit hose in order to hook up the hose extension prior to egressing through the upper hatch. By this time I was really hot pouring sweat. The capsule was very hot after reentry and I really noticed the increase in humidity after the snorkels opened. I decided that the best thing at that point was to come out the side rather than through the top. I am sure I could have come out the top if I had had to, but I did not see any reason to keep working to come out the top. So I called the ship and asked them to clear the area outside the hatch. When I received word that the area was clear, I removed the capsule pin and hit the plunger with the back of my hand. It sprung back and cut my knuckles slightly through the glove. The noise of the hatch report was good and loud but not uncomfortable. In summary, my condition is excellent. I am in good shape; no problems at all. The ASCS problems were the biggest I encountered on the flight. Weightlessness was no problem. I think the fact that I could take over and show that a pilot can control the capsule manually, using the different control modes, satisfied me most. The greatest dissatisfaction I think I feel was the fact that I did not get to accomplish all the other things that I wanted to do. The ASCS problem overrode everything else.

* * *

MA-6/13 DEBRIEFING
ASTRONAUT COLONEL JOHN H. GLENN

February 20, 1961

A. Prelaunch

A1. Were there any problems with the insertion and countdown procedures?

I thought in general, things went pretty smoothly during the prelaunch phase. I got all the information I really needed. The actual insertion itself went very smoothly. We've been through this a number of times and Joe Schmitt knows exactly which strap to pick up and when. He does a wonderful job of getting you plugged into the capsule. So there was no problem from that standpoint at all. The practice that we had had before I think stood us in pretty good stead for a smooth insertion because we find we're working as a team. When Joe reaches for a certain strap I just automatically turn my body to let him pull it out to a certain point and when he's tightening it up, I know just where to squeeze in a little bit to help him. I guess maybe it sounds rather peculiar but even something like that becomes sort of a team maneuver. We have been through this so many times that it was very easy and went very smoothly I thought. We had some problems of course, such as the problem with the mike in the helmet that delayed us a little bit. I guess we have probably pushed those mikes up and down many thousand times with no trouble and it had to pick that particular time to break, right in the middle of a count. It was the little fitting that slides up and down on the left mike that broke. Joe was able to change it rapidly. We had problems with the hatch bolt also, which delayed us for a little while. But this was no problem as far as I was concerned. I was surprised at the leak rate we achieved. All they would read it as was below 500 and this is phenomenal. The best we have ever registered before on capsule 13, as I recall, was 610. So this was excellent. I think that the constant bleed was probably greater than the leak rate. Later on we did build up a pressure slightly above ambient in the capsule. The card storage for the star charts was not ideal at all, even for getting them in and out on the pad. We need some better way of stowing things. The map book storage just doesn't seem to bc too good either; you shove that map book in and out of that case trying to review things and leafing through it is very awkward. If the pages are open, this adds to the difficulty. I don't know how we are going to improve it. I felt that it was unnecessary to go back over the switch positions a second and third time during the count. It seemed to me that the last time we go over the switches and set the fuses is completely redundant and unnecessary, a sort of WPA [Works Projects Administration, depression-era] program designed to keep the astronaut busy. I was not particularly annoyed. I didn't have anything else better to do at the time; but to have something in there just to keep the astronaut busy isn't necessary.

A2. Were your communications and information flow satisfactory at all times during the countdown?

I thought Scott [Carpenter] kept me informed very well without a lot of extra chatter. He gave me all the information I needed to know. The

MOPIS circuits and our cutting in and out on communications worked very well. We had good volume control up and down; all I had to do was say that something was too loud and the volume was controlled adequately. I thought all the communications went very well. The RF checks all went better than I had ever heard them go before they were all very clear. I think the only time anybody didn't hear another station on the RF was when Rescue 1 had a little trouble getting the blockhouse. I came up to make sure they could hear me and they reported that could hear the capsule OK, and then they picked up the blockhouse immediately thereafter.

A3. Did you have any problems with pressure points, stiffness, ventilation, etc., during the countdown?

I had no particular pressure points out of the couch, no stiffness. It seemed that after I had been on my back for several hours I had a spot about across the shoulder blades that, while it's not extremely tender, is an area you realize has more pressure on it than any other part of your back. I remember after the five-hour session up there on my back the day we scrubbed, the next day I had a little spot across the same section of my back that felt a little different but that is not a real pressure point. The ventilization was adequate. I was comfortable. The temperature during the launch count was cool and I intentionally wanted to cool on down as much as I could at that time. Along about the last 20 minutes of the count I was beginning to get too cool. By the time I started calling for them to warm things up a little bit, they couldn't change the freon flow; I had waited too long on that one. So I just exercised two or three times to pep things up a little bit and then I was very comfortable.

A4. Comment on the length of the prelaunch period.

I do not know exactly how long we were up there this time prior to launch. As I recall I got in the capsule about 6:05, was on my back, back in the capsule, from then until we launched at 9:47, a total of 3 hours and 42 minutes. There wasn't any problem with this time period. The shorter we can keep it the better, of course, but I think I was still in very good condition and was not fatigued to an extent that would affect any of my operations.

A5. Do you feel you had sufficient training in prelaunch operations directly involving you?

Yes, after all the counts we have been through, all the practices on the pad, the simulated flights, and the launch simulations, I think I have been adequately trained.

A6. Was the ground checkout procedure with the Control Center Capsule Communicator and other MCC operators satisfactory? Were you satisfied that MCC operators were happy with the T/M measurements at lift-off?

Yes, if we were to repeat this flight right now, I cannot think of anything I would want to change in the checkout procedure with the Control Center.

The fellow in the capsule does not particularly want to get involved in all the workings of the Control Center. He doesn't need to. And I would just as soon keep all extraneous things that you're not particularly interested in off of the circuits to the capsule. Go ahead and check communications and let the fellow in the capsule have as much spare time as you can. Don't give him a make-work program because he has a lot of things he would like to review such as the checklists on the panel and the flight data cards.

B. Launch and Powered Flight

B1. What were your predominant sensations during powered flight and relate these sensations to the over-all flight environment and to your previous piloting experience if possible?

Well, in particular, the vibrations, some of which we have run into before. One that comes to mind is the sensation of shaking due to the limber spring effect of the booster. It is very similar to an F8U [aircraft] in rough weather. The cockpit of an F8U sits out so far ahead of the rest of the airplane that it is a limber affair. In rough weather, you have a feeling that the cockpit is bouncing up and down quite a bit. It was the same type of sensation that you get on the At1as.

B2. Did the ECS perform properly during the powered flight phase?

Yes. I had no problem with ECS at all during the powered flight phase. Temperature was comfortable. No malfunctions at all with ECS during that phase.

B3. Did you notice any vibration of the escape tower at lift-off or near maximum g?

You cannot really pin down where the vibration is coming from. I don't think you can differentiate between one type of vibration at that time and another. You have the sensation that the spacecraft is vibrating but it's not possible to determine whether it is from the tower.

B4. Did vibrations interfere with the readability of any of the instruments?

No. I don't think they ever reached a severity where I had any trouble reading the instruments.

B5. Did any capsule components vibrate excessively during powered flight? Describe.

See section describing sensations during launch and powered flight.

B6. Did all telelights operate correctly, through separation? If not, what did you do?

Yes, all telelights did operate correctly through separation.

B7. Comment on the quality of the voice communications during the powered flight phase.

Voice communications, I thought were very good. There were a couple of times, I believe, when Al [Shepard] and I blocked each other on a transmission. That was probably my fault. I was trying to get as many impressions back to the ground by radio as I could. Perhaps I overdid it, and I tried to get 'too much back. The quality of voice communications was good and I do not believe there were any communications problems.

B8. Can you identify the major source of background noises at various times? (booster, aerodynamic, RF, inverters, fans, etc.)

Well, to a degree, you can. During the high g period, I think you can hear a rushing or aerodynamic sound, like rushing wind. I think you hear that as a separate noise from the thundering noises you hear from the booster. The noise of the inverters and fans and the high-pitched 400 cycle tone, that we're so used to, are there all the time. You can not get away from that.

B9. Did you think the tower either hung up or struck the antenna fairing during separation?

No. You feel a bump when it leaves but I think it's mainly a change in acceleration. I think it pretty well matches the g profile, which shows a little bump when the tower leaves. You can feel that you are unloading about 900 pounds there in a split second. I did not feel that the tower hung up or struck the antenna. However, it happens so rapidly, it is gone by the time you see it start firing, that I can not be really sure whether it struck anything or not.

B10. Did the escape motor shake the capsule or make the shingles flutter?

No. The only affect on the capsule that I noticed was that it left a coating on the window. This was not serious. Up to that time. The window had been nice and clean, but the tower jettisoning left the window coated with a little film. This was very noticeable later in orbit. When the light reflected off the coating at certain angles, it made it difficult to see through the window. Later on in the flight I noticed that there was a long, thin, red streak down about two-thirds of the outside of the window. The red streak ran down the right side of the window as though a bug had been smashed against it. I remember wondering if we ran into some bugs during launch.

B11. Did you see the tower during separation and did you see the exhaust? For how long could you observe the tower following separation?

I think we already covered that earlier.

B12. Are you satisfied with the procedure for reporting cabin and suit pressure during the critical phase from 80,000 feet to 100,000 feet from both a "ground" and your point of view?

Yes, I believe the ground received its information right on time. I was watching the cabin pressure and reported it when it was sealed off. It stopped first at about 6.1 psi and bled on down very slowly. I remember being a little bit surprised that it stopped as high as it did. Later on, during the flight the thing had come down to an indicated 5.5 and held very steady, it was just on solid five five.

B13. Did you note the pitch down of the booster at the time of booster cutoff through your window?

I did not notice it as a rate that I would sense. I didn't feel that I was suddenly being pitched down. But I certainly noticed it when the horizon view came up on the window. It came up on the window, right at the same time that I was counting the seconds off to myself and watching for the tower to go.

B14. Could you discern booster cutoff and tailoff characteristics? By which of your senses?

You can certainly feel the g changes that accompany the booster tailoff. You feel the g's by your body senses. There is also at booster cutoff a change in vibration because two engines are dropped. I think you also sense a change in sound as a result of jettisoning the booster engines.

C. Zero-g Phase

C1. Could you sense separation of the capsule clamp ring? How? Did capsule separation occur immediately? How did you tell?

There is no doubt about separation of the capsule clamp ring. You cannot tell whether it separates except by the sequence light, but you certainly can tell when the clamp ring goes out because you hear it. There is a very definite report when the clamp ring bolts fire. The capsule separation did occur immediately. I was mildly surprised by the force of the posigrade rockets. I had thought that the spacecraft would appear to just drift off the booster and that you might not even be aware that the posigrades had really fired. But there is no doubt about it. They give you a definite little jolt when they go off. You can also hear the posigrades fire just as you hear the retrogrades fire later on.

C2. Was the booster pitching, yawing, or rolling at this time?

I was surprised that just before SECO, it felt like I was out on the end of a big, long, rather limber spring. Apparently when all the LOx is out of the tanks, the Atlas has a lot more give and swing to it. The motion was random, it was not limited to one plane. The thought that came to my mind was that the ASIS might kick off, but it didn't. The booster was not rolling. The oscillation was in pitch and yaw. As you approach SECO, there seems to be a build-up in the vibration, which I experienced, earlier in powered flight. I imagine this is probably due to an increase in engine vibration at that time.

C3. Could you detect posigrade rocket firing? Describe.

Yes.

C4. What motions did the capsule go through at separation? What were your cues?

I think it came straight off the booster. I don't think we had much random motion because there wasn't any damping action. My only sensation coming off the booster was of a very slight pitch down, as though my head had gone over toward my heels. This is a normal sensation from a reduction in linear acceleration. It is possible that we might have pitched down just enough to give me that sensation, but I don't think we did because I was watching the rate indicators and I don't recall any real pitch down motion.

C5. Comment on ECS cooling during weightlessness.

The comfort control calibrations from the chamber runs at the cape are not accurate at all apparently. The setting of the suit circuit comfort control, which provided cooling at the Cape during our ECS runs was marked in red, right in the middle of the scales. By the end of the flight, I had increased this water flow on the suit circuit clear past the maximum setting on the scale, which was 1.7 pounds per hour. My setting was beyond that and the steam exhaust temperature was still only down to about 470. I increased it slowly even beyond that point. I did not want to turn it way up for fear I would freeze the outlet up, so I kept edging it up slowly until I finally wound up with it clear above scale. On the other hand, the cabin temperature got up as high as 105^0 at one time. I increased the flow slowly over a period of a full orbit and the cabin temperature gradually came down to about 95^0. When I tried to increase it even more to get the temperature down below 95^0, the excess water light came on, so I had to back it off. The water light went off in about 10 minutes, so I turned up the water again. I turned the water off every time as we were suppose to prevent hysteresis in the valve. When I would bring it back up again it would freeze. The excess water warning light was on and off 5 or 6 times. I never really achieved one setting, which was close to the limit, and still low enough to keep the excess water light off. I just didn't have a good feel for the cabin water position at all. It froze up rapidly on the position that was satisfactory in the chamber at the Cape. Thus, the planned setting on the suit circuit determined at the Cape is too low, while on the cabin circuit the planned setting was too high.

C6. Did the ASCS sequence properly and did it hold the correct attitude (rate damping, turnaround, etc.)?

During rate damping, turnaround worked properly. It worked properly all during the first orbit. Everything went according to plan during the first orbit. We had that long period on ASCS in order to hold an optimum attitude for radar and communications. Just as I got to the coast of Mexico, the capsule started drifting to the right in yaw and it would drift over to about 20 degrees, instead of the normal 30-degree limit, and then the high thruster would kick on and bat it back over to the· left. It would overshoot to the left and then it would hunt and settle down again somewhere around zero. The spacecraft would then drift again to the right and do the same thing repeatedly. Sometimes it would do this just in yaw but at other times when it kicked into orientation mode in yaw, it would drop into orientation mode and all three axes would go through a period of fluctuation and the whole capsule was doing this until it gradually settled down. I let it do this 3 or 4 times and decided it was just too wasteful in fuel to let the thing continue, so every time the yaw would kick in, I would take over manually, recenter it slowly, and reengage ASCS. It had that same characteristic every time I turned the ASCS back on for about some 15 or 20 minutes. On the second orbit, when I was pretty well over toward Africa, it reversed, and then it would drift to the left in yaw instead of the right. During the difficulty with yaw, when I would try to use low thrust to bring it back, there was no left low thrust. That's the reason the spacecraft would drift off course and hit orientation mode; where the high thrust had to take over. It seemed that later on if I held an attitude other than just zero-zero and -340 for any period of' time at all, any axis was liable to drift off in any direction and I never could pin this down to a pattern. This is what consumed the bulk of my time for the last two orbits. I was trying to analyze the trouble with the ASCS, so we would know exactly what it was doing but I was unable to pin this down to any exact pattern. The attitude indicators were most in error after I yawed around 180 degrees and held this attitude for several minutes, looking at these little bright spots at sunrise. I had had the capsule lined up in orbit attitude just before I made the 180^0 turn. I had caged and then uncaged the gyros after visually aligning the capsule to orbit attitude by reference to the horizon. Following my return to orbit attitude, the indicators were in error by 300 in right roll,

35 in right yaw, and 420 pitch up, which puts it some degrees higher in pitch than it should have been at that point. I don't know whether this got reported or not. I felt that the gyros should have been aligned properly when I returned to orbit attitude unless there was some malfunction. During the majority of the first orbit, all the control systems functioned perfectly. The ASCS performed the turnaround maneuver properly and dropped into orbit mode. After Bermuda, I started into the controls check that we had practiced so many times. The test went just like clockwork. It was just like running it on the trainer again; the same rates and the same reaction of controls. I was really elated by the precision with which the controls check proceeded. It is quite an intricate little check until you practice it a number of' times, and there are quite a number of switch and pull handle activations during this check. We had run through this so many times at the Cape, it was really a pleasure to see this thing work properly. Thus, the ASCS was working properly at that time.

Following the controls check, I went back on ASCS and it held fine all through the first orbit. I could not pin down any pattern to the malfunction of the ASCS. I tried caging the gyros and getting everything lined up and then uncaging the gyros. Then I would try flying on the fly-by-wire to see if there was any connection between that system and the ASCS problem. It would still drift off in random fashion. So I would recage again, uncage, and try the same thing again on manual. There was just no pattern that could be established to it. Whatever the problem was, it was random.

The fly-by-wire system is excellent. That's one that really gives you fine control. On fly-by-wire you can control right to zero rates with no problem at all. I liked it very much. The manual system was not quite as good as on the procedures trainer. It felt mushy. I think this had been predicted from the tests that were run on capsule 15. On manual, I didn't feel I had as accurate control as I did with fly-by-wire. The low thrust fly-by-wire is excellent for very accurate control when it works. Yaw was the only dimension that had any problem. There was never any problem on fly-by-wire in roll or pitch.

C7. Could you hear operation of the control system (i.e., firing of the control rockets, action of solenoids, control linkages, etc.)?

No. More than hear them, I think you sort of feel them as a vibration not as a movement of the capsule, but as the thruster operates, you feel a very high frequency buzz. I think you feel this more than hear it. You can hear the high thrusters. They are not very loud but you can hear them. But the low thrusters; you feel them more than actually hear their operation. Yes, you can hear the solenoids click, though I don't think you hear all of them. You can hear some of them clicking and I was also picking up some of the solenoids in the headset, too. Evidently, the signals the solenoids were making feed through into the communication system.

C7-B: Could you see the exhaust from the pitchdown thrusters?

A couple of times I saw the low thrust pitchdown thrusters. This occurred when the light was just right. In the case of the high thrusters, any time that the light is coming from behind the capsule, you could, see the exhaust if you were looking against a dark background. If you have the nose of the capsule down where you have a bright land or cloud area as a background then you cannot see it at all. This was very noticeable at the sunrise period when you still had a dark sky and a dark background of the Earth toward the western horizon that silhouetted the steam coming out from the thruster. The steam from the thruster reflected the brilliant sunlight through the window. You could see a fan-shaped cloud of vapor 20-wide coming out of the thruster. The fan shape of the vapor was very clear cut, very sharp, and even.

C8. Comment on the quality of the hand controller for each manual control tried (ie., effectiveness, backlash, slop, binding, lag).

I didn't feel that there was any particular trouble with the hand controller at all. It was effective. I think that the fly-by-wire switches had been reset to about where they had been before. They felt very similar to the procedures trainer. There was no binding in the system at any time. It did not tend to hang up at all. I think it was correct the way it was finally rigged.

C9. How did the performance of the manual control systems compare with what you expected as a result of training on the various Mercury simulators?

The procedures trainer is a very close simulation. I think the manual system on the procedures trainer is a lot more crisp and gives a lot better result, particularly at low thrust, than the capsule did. Fly-by-wire action is very similar, in fact, identical. On the ALFA trainer, I think the characteristics of the control response are probably very close, but I somewhat revised my opinion of the ALFA trainer as a result of this flight. Whether we realize it or not, we sense and react a lot more to our body movements on the ALFA trainer then we realize. This may not be true for other people, but I think that I at least react a lot to my body movements on the ALFA trainer, as well as to the instruments I am looking at. Now in the actual capsule, once you get into orbit, you just don't have these same body sensations at all. They are just not there. You go by what you see. And, in that respect, I think that the procedures trainer actually comes closer to the type of presentation that you have in orbit.

C10. Did you have any unusual physical sensations from capsule motions?

No. That's exactly the point. I don't think you really have much· of a sensation of movement at the rates that we are talking about. You go by what you see your attitude is and react to that. In flight, there are no motion sensations to which you can react. I think in the procedures trainer, you just ignore your body sensations and. react strictly to what you see. On the ALFA trainer, you tend to combine the two and react not only to what you see but what you feel.

C11. Could you feel angular accelerations during capsule motions?

Only at the higher rates. If you get up to a rate of about 5 degrees per second, you are usually accelerating rapidly enough in achieving that rate that you can feel the acceleration. Most of your rates, of course, are much lower than that, down around zero to 2 degrees per second. Thus, during most of your maneuvering you just don' feel much in the way of acceleration and deceleration when you start and stop.

C12. Did the periscope, window, and instrument attitude references agree during the mission?

Periscope and window did. Instruments varied. (See Question C6)

C13. Could you determine yaw attitude: (a) By checkpoints through the window? (b) By checkpoints through the periscope? (c) By terrain drift through the window? (d) By terrain drift through the periscope? (e) By star drift through the window?

I believe there was a period of learning to use the periscope and window during the flight. I think I wound up at the end of the flight much better able to discern yaw and set the capsule on the correct heading than I was at the beginning of the flight. Roll and pitch are so definite, that there is no doubt about controlling the attitude in these dimensions. You sit there and just drive the window where you want it. With accurate control on fly-by-wire it, is very easy to position the capsule exactly where you want it. But this wasn't quite so true in yaw. Your reference in yaw is not as good. I found that the best way I could set up yaw was looking out the window and aligning with some object that was going straight away from me. The procedure that seemed to aid this more than anything was pitching down. If I would pitch down to about 60 degrees or so where you have a pretty fair vertical view, then I could pick up clouds and land moving out from under me at a much more rapid rate than in orbit attitude in which I was looking way off toward the horizon. When you are viewing the horizon the land marks are much farther out and their apparent motion is considerably less if you are pitched down and are looking at them coming right out under your nose. When they are doing that, it's pretty easy to see what your yaw is. It's much more apparent that things are moving at an angle across the window when they are going at a greater speed. I wound up pitching down whenever I wanted to set yaw toward the end of the flight. The periscope I think just required a little more practice. At the beginning of the flight, it was a little bit difficult to pick up yaw rate, but I think by the end of the flight, I could probably set yaw on the periscope down to within a couple of degrees. I would say that this is about the tolerance of accuracy on the periscope; I don't think I could have got any closer on that. I think using the window you can get down to about the same accuracy maybe, possibly a little better. I think the big difference between the window and the periscope is probably the speed with which you can determine yaw. Looking out the window, you just come right around until it looks right, that's it, and it" very apparent to you because you are looking directly at the clouds and they are moving rapidly. On the periscope I seem to have to do a lot more maneuvering and look and watch and track a cloud along the line and then make an adjustment, then track another cloud along the line and so on. It took longer to do it on the periscope than it did looking out the window. You could determine yaw rate by star drift through the window. This took a little bit longer. If you were aligned pretty much on track it seemed a little bit more difficult to track the star and make your final accurate alignments in yaw. In fact, without watching it for a very, very long time it's apparent if you turned a little bit yawed one way or the other the drifting of the star field becomes more apparent. If you pick a particular star and track it, and it drifts rapidly while your rate indicator shows that the spacecraft is not yawing it is apparent that you are not quite on track. But, once you get this aligned pretty well on track, adjusting within the last 10 degrees of yaw is fairly difficult at night. The periscope gave no help at all in aligning with anything at night. You could see different patterns in the scope at night because we had a full moon on these cloud decks but this wasn't much help it's so dim through the periscope you couldn't really align with any particular spot and follow it as a yaw reference. So your yaw reference at night, if your instruments are not operating properly, is strictly out the window. You use the window in conjunction with the rate indicator when the yaw attitude is widely displaced from track. If you track a particular star since you have no sensation of rate you must cross reference to yaw rate indicator to be sure that the motion you see is not due to a capsule yaw rate. You have no sensation of turning unless you do look at the rate needle. Your rate indicators operate properly at all times. Aligning at night in yaw is not very easy. You can see cloud patterns out the window and that helps some. With the bright moonlight we had you could keep aligned by reference out the window on some particularly outstanding cloud formations. This was much more rapid than star reference. I think you could probably align out the window, almost as rapidly as you could in the daytime, when you have a full Moon which illuminates the cloud formations.

C14. Could you determine yaw:

(a) By movement in the periscope?

(b) By movement in the window?

See Question C13.

C15. Did sunlight effect the visibility of any of the instruments or controls? Was it necessary to adjust the capsule light to compensate for this effect?

The sunlight coming in the window is very brilliant and intense, extremely bright, very clear brilliant white light. It reminds me of the pad with the arc lights on. One of them shining in the window is a brilliant white light that's just what it's like. Just like someone had an arc light right outside with the brilliant white light they put out. I didn't change the capsule lights for this at all. We had thought before the flight that we would probably not be able to look directly at the Sun. I sort of peeked around the edges of the window to try looking directly at the Sun without my eyes being protected. Previously, I had used the filter that we had for this purpose in the open position. By squinting the eyes it was possible to look directly at the sun with no ill after effects at all. This would be very similar to looking directly at the Sun from the Earth. You don't accomplish anything by doing it since you can't see anything anyway with your eyes squinted down that far. But it is interesting at least that the Sun is not quite as intense in that regard as I thought it would be. It's not completely blinding if you happen to open your eyes to the Sun. I know a number of times when I'd be maneuvering the capsule, the Sun would come across the window and by squinting my eyes and looking to another part of the capsule, I could still see the instruments and see what I was doing, even though the Sun was coming directly in. It's very bright, I don't want to minimize it, but it does not stop all other activity when it happens to shine close to your face like, we thought it might. As far as the lighting of the capsule goes, the fingertip lights were very useful. We had thought from some of our practice in the Procedures Trainer the little finger tip lights we had might be very helpful and they were. I used them numerous times to look at the charts. They were also useful before my night adaptation was complete, for making gage readings or trying to read out a report. The instrument panel is not evenly lighted in the capsule. You really need edge lighting on some of the gages to make them readable. We have flat red lighting on gages that are rather far away from the light sources, so that the illumination is dim. Another factor in capsule lighting is that you need to be dark adapted to where you can really make the maximum observation of the sky shortly after sunset, or determine yaw position from a cloud formation out the window immediately after sunset. To do this, you would have to adapt your eyes a lot more in advance of sunset than we had planned to do on this flight. I tried for maybe a minute or so before sunset we had an eye patch and I tried it on the first orbit. It was sort of a jury rig item that we put together at the last minute with tape around the edges to hold it in place. The tape just didn't stick the way it should have. It kept coming loose and I kept changing its position. I thought from our previous practice that it would work all right, but it wasn't worth much. I gave up on it and stuck it over on the emergency O_2 rate box. It wasn't much good. So I tried then just to shut my eye and adapt one eye while we were coming up to the dark side so that I'd have pretty good reference after we went through sunset. I just wasn't trying it far enough ahead because to make any real observations or really pick up yaw reference rapidly, you have to be night adapted farther ahead than I was.

C16. Describe the appearance of the Earth relative to colors and light intensities. (Refer to Earth-sky camera pictures.)

Looking back at the Earth, colors and light intensities were the same as flying at high-altitude in an airplane. You look down at the ground and you see the same colors and the ground looks the same as it does flying at 40,000 to 50,000 feet in an airplane. Now if you look off to the horizon, the view is completely different, of course. It is also different if you look up at the sky. But looking back at the ground it's just like you were in a high altitude airplane of some kind. I could see patterns on the ground and rivers. I particularly recall the area around El Paso, as one of which we had made a photographic study. The green irrigated area that goes up and down the valley each side of El Paso can be easily seen. It contrasted well with the desert area of each side of it. You could see squares of the irrigated land. I won't imagine that I was seeing the individual little irrigated plots that are only a few hundred yards long. I don't know what these were; maybe they were quarter sections. I could see rivers and lakes. You can see several cities, such as Savannah and Charleston, very clearly.

Knowing that there is a river winding through an area, you interpret little blobs that you see as being bridges across the river. I think you can probably pick out areas down to maybe 100 yards long. We were figuring before that just using a gunsight on a fighter plane, you could pick out areas down to one mil. I think we figured that 1 mil at this altitude would be on the order of 100 150 yards. On the night side the visibility is pretty much dependent on whether the Moon is up or not. In one area off the east coast of Africa before the Moon was up I could see nothing in the way of cloud formations or anything at all. Reference then was strictly by the stars, but even without the Moon being up it didn't appear to be any problem at all seeing the horizon because the stars were so clear. Maybe you do have some very low intensity light that comes back off haze in the atmosphere that aids in seeing the night horizon: When there is no Moon out, looking right straight down this is like looking into a black hole. We had wondered about whether you could see lightning flashes at night and they were very visible. There were two big storm areas, one big general storm area that was just north of track off the east coast of Africa and another smaller storm center that was south of track and a little bit east of the first one. The lightning flashes in both of these storm areas were very visible. You could see the lightning going from cloud top to cloud top sort of sheet lightning effect. You could see lightning in the clouds lightning of the type you normally associated with clouds of vertical development where there is one single big flash rather than going from cloud to cloud. These were very visible. I tried the air glow filter in looking at that area but it was so soon after sunset that I don't think I was night adapted enough to really pick up air glow. The Indian Ocean ship said they set their flares off, but I did not see them.

C17. Were the controls easier or harder to reach under zero-g?

No difference. I don't feel there is any difference in reaching for controls or controlling switches or knobs under zero-g than there is at any other time. I think that most of us who have flown quite a bit have gotten use to being at odd-ball accelerations. You consciously reach for something and control your hand until you get to it. It's the same thing under zero-g. There was no tendency to come out and overshoot a switch and have to put your hand down. I just never had that problem happen right from the start of the weightless period. Just reach for what you want and that's it. I think you just glance at the switch that you want to go to and you control your hand visually all the way out. I don't think you put your hand out by feel like this when you are looking at a knob. You put your hand out to go to a certain knob and whatever force you need on the muscles of your arm to get there, that's what you use. I think perhaps there were some erroneous impressions from some of the studies that Wright Field was running. We were subjects on some of them because we happened to be onboard for training. Changing from a 2-g pull up through a period of weightlessness and back into 2 g's again, there's bound to be some overreaching and underreaching. But where you're in a constant g state, even though it is zero-g, you just compensate for this automatically. If I wanted to reach for something or go back to controlling attitude, just for a moment, rather than put the camera away, I'd just leave it floating where I had used it. It's something to laugh about now, but it just seems natural up there. I think that the human is more adaptive than anybody thought. It wasn't something that I had thought of before and planned to do; it is just that you put it there and you know it will stay. You don't need to think anything about it. I noticed as I moved the capsule that things would drift. Some of the things didn't stick on the velcro and when I would maneuver and stop a rate, something would come drifting across.

C18. What items, if any, vibrated during zero-g? Did you consider any of the vibrations annoying, excessive, or possibly damaging to any capsule system?

None that I know of.

C19. Can you identify the major sources of background noise at various times during the zero-g period? (RF, inverters, fans, tape recorders, cameras)

No, this was pretty constant. You have a little different type of background noise on HF and UHF. The inverter, fans, tape recorders., cameras, were operating all the time and this was a pretty constant noise level.

C20. Was the noise level annoying?

Well, I think we have probably become pretty well acclimated to this through a great deal of time in the capsule. It's not desirable, that's for sure. But it's certainly no worse than you have flying in a fighter airplane.

It certainly is not annoying to the extent that it interferes with your activities. Any noise of this intensity is undesirable but it does not interfere with the operation.

C21. Were there any floating objects in the cabin during weightless flight?

Surprisingly few. I really expected to see more things floating around. In particular I expected to see a lot more very small flying particles; a dust and a little fillings and things like that, but there weren't too many of these. The capsule had been pretty well cleaned out. There were a few washers and I picked a few little odds and ends out of the air. We had put up some double-sided tape over on the hatch so we would have something to which to anchor floating objects. I took some of the floating objects and stuck them over there. One of the washers came off the tape and I noticed an orbit later this little washer came floating back. I got it again and looked over where I had put it and it wasn't there, so I think it was the same one. The only two items I couldn't control were two 1/8" inch bolts that were loose inside the periscope. They kept floating in and out of the view. Overall the cabin was surprisingly clean, especially after some of the pictures we've seen of previous flights where there was all kinds of litter floating around.

C22. What did you see, hear, and feel during retrorocket firing? Describe.

Apparently the alignment of the retrorockets was such that it initially displaced the capsule in yaw, because the first motion I recall was when the yaw rate went full-scale to the right. I brought it back with manual control. I don't know whether ASCS was helping. I suppose it was, although the ASCS yaw operation had not been reliable at all, to say the least. But yaw came right back into the center and held in the center during the rest of the retrorocket firing. You do hear the retrorockets firing very definitely and you certainly do feel them. Al [Shepard Jr] and Gus [Grissom] had both noted previously that there is no doubt when the retros fire. This is a very positive bang on the back at that time. It occurs very rapidly and very sharply. The sensation was interesting during retrorocket firing; you're sitting there weightless and although you know you're going very rapidly towards your back, when the retrorockets fire, I had a sensation as though I was being accelerated back towards Hawaii. The very same acceleration feeling that you feel during launch. It doesn't take much effort to kill this sensation, when you look back out the window at the end of retrorocket firing and see the ground still going in the same direction. Nevertheless that was the sensation you experience when you're concentrating on the instruments during retrofire.

C23. Could firing of individual retrorockets be detected?

Very definitely. Each one hits you with about the same additional force. There is no doubt about there being more deceleration during the period when two of them were firing. The cutoff was not as positive as the initiation of the firing. There is no doubt about when you go on to second one, as when you ramp down to the third.

C24. Could you sense separation of the retropackage and did it affect the capsule in any way?

We retained the retropackage so it did not jettison during the weightless condition. (See reentry section)

C25. Did retropackage straps spring back and strike the capsule.

See reentry section

C26. Compare the sensations (sight, sound, feel) during retrofire with those experienced in the ALFA Trainer, the Centrifuge, and the Procedures Trainer.

That's sort of a tough one because I don't think any of the simulations we have had really reproduced retrofire. Now the movements we expect from the rate needles, these have been pretty well duplicated on the Procedures Trainer and the air bearing, but not the sensations. The centrifuge doesn't do it because the centrifuge has to build up to their speed changes much more than the actual retrofire does. I just do not think there has been any trainer that really simulated retrofire. On the ALFA the motions are the same as far as controlling goes but I think that the most noticeable thing is the changing g field. After you have been there under zero-g for such a long period of time, the changing g field hits. The retrojet actions are all very positive; there is just nothing that we have done that is similar to this.

C27. If you controlled retrofire manually, what was the difference in the degree of difficulty in controlling actual retrofire as compared to controlling the retrofire simulations presented to you during your training on the ALFA Trainer, the Centrifuge, and the Procedures Trainers?

Well, we were set up to do an ASCS retrofire. The ASCS was working okay at retrofire time. I backed it up manually. I don't know whether our problem might have been in horizon scanners during this flight or what; but during the period just prior to retrofire, this ten minute period when the scanners are cut on and we have slaving so that we are correcting, it, appeared that everything corrected right up. The window periscope, and the instruments all matched perfectly as we came up to retrofire so I left it on ASCS and backed it up manually. I played it as though the ASCS wasn't working at all. So we really had ASCS and manual operating at the same time. This is not a mode that I had practiced particularly in the Procedures Trainer, so it's a little difficult to compare it with the Procedures Trainer runs. It was very effective. I would guess that none of the axes varied more than 3 degrees at maximum from their proper setting.

C28. Did the whole-body-motion training you received on the ALFA Trainer help you in flying the capsule at zero-g?

See Question C26

C29. Were you aware of any resemblance between the angular acceleration cues you experienced during weightlessness with those you experienced on the ALFA Trainer?

See Question C26

C30. Did you see the booster after turnaround? For how long and describe what you saw. Did you see it through the periscope when it passed below you?

Well, that was a real good sight. As it came to almost its proper position, the capsule was pitched up a little bit and here was the booster right in the middle of the window. It wasn't more than 100 to 150 yards away. That was some sight, this great big booster sitting right there, just that far away, and going away, thank goodness. It was sitting at an angle with the capsule into the booster pointed down maybe 30 degrees and the longitudinal axis of the booster angle may be 45 degrees to my position so that it was pointed down and northeast. As the capsule corrected back then settled, pitched down a little bit into orbit attitude, the booster went out of sight at the top of the window. I could duck my head down, though and look up and see the booster. I have to give Carl Huss credit. He drew up the chart on exactly where the booster would be and he couldn't have hit it better if he had been there I don't think. I watched, off and on for some 6 or 7 minutes. It got farther away and was drifting lower. The last time I saw it, it was probably a mile behind and a half-mile below. I looked for it again in the periscope, over Kano, where he had predicted it would be, directly underneath and just a little bit south, but I couldn't see it then. You wonder what your capabilities are gonna be. I had good control of the spacecraft and I reoriented to orbit attitude, put it back on ASCS and didn't drop into orientation mode or anything. I think you could have come off that booster with maybe running a controls check, that plus maybe another 5 minutes of just maneuvering to get the feel of this thing. I think you could have controlled accurately enough right at that stage to have gone back over and probably made an attempt to join up on something.

D. Reentry

D1. What control mode was used during reentry? Comment on effectiveness.

I went to manual control after retrofire because I set my roll rate in at about the time we had reached a quarter- or a half-g. The early part of the damping was okay. I didn't feel that I had all the authority that I wanted when the oscillations started building up and I went to fly-by-wire in addition to the manual. I don't think this was until after .05g. I'd like to reserve comment on that until we look back to the record because I'm not real positive where I went to flyby wire in addition to manual.

D2. Compare your reactions to the reentry acceleration profile to corresponding reactions experienced on the Centrifuge.

The g's were no problem. This is similar to centrifuge runs as far as the g tolerance goes. You could still talk at 8g's and although you apparently didn't hear me, I was communicating through that area of g buildup through the communications blackout period. We had all been able to communicate on the centrifuge up to varying levels up to 14g's, although at 14 they were pretty much grunts and some largely unintelligible things. But there was limited communication up to about l4g's so communicating at 8g's was no particular problem. I think our experience on the centrifuge was very similar to this as far as acceleration goes.

D3. Comment on ECS cooling during reentry.

I never could get the ECS cooling system pinned down to where I felt I had very accurate control of it. I was comfortable, but I would like to have had it a little cooler all during the flight. There was no doubt when the heat pulse occurred during reentry. I had the window shutters open and could see the glow outside. The ECS cooling was okay during the early parts of reentry and I didn't feel particularly hot during the heat pulse. However, about the time we were getting down to around 75,000 or 80,000 feet it got very noticeably warm. It went from the same comfortable level I had been used to up to where it was uncomfortably warm in about 15 to 20 seconds and I never had been able to cool the capsule down as much as I wanted so I was not precooled when starting into the reentry. The capsule stayed uncomfortably warm all the way; down through drogue and main chute. When I was on the water finally I was sweating profusely.

D4. Were cabin or suit pressure changes excessive during reentry?

This question was not answered.

D5. Did you lose communications during reentry? For how long?

Yes, I lost communications during reentry. I kept communicating, however, to get some sort of record of how long we are blacked out and I don't know how long it was.

D6. Did any capsule components vibrate excessively during reentry? Could you pinpoint time of occurrence?

No, you have sort of a general vibration and an increase in noise during that period. I wouldn't say that anything vibrated excessively.

D7. What was your first cue of g reoccurring upon reentry?

There is no doubt when you start coming back into the g field. You start settling back into the couch again. During the weightless period the natural position is to be slumped forward with the head about three inches out of the couch instead of being back in the couch as for proper eye position.

D8. Were there any oscillations of the capsule during reentry? Could you estimate their amplitudes? Were they bothersome to you in a physical sense?

Yes, there were oscillations through the main g pulse and these were easily damped out. Following the main g pulse there were just minor oscillations. When we got down to about 50 or 55 thousand feet the oscillations were divergent and they built up to an estimated plus and minus 70 to 80 degrees. The oscillations were diverging so rapidly that I was afraid I was going to get completely over, so that if we fired the drogue it might get wrapped up in the capsule. I was going for the manual drogue button at about 35 thousand feet, and as I was reaching for the button the drogue came out by itself, at about 30 to 35 thousand feet. The frequency of oscillation prior to drogue was about one cps [cycles per second]. As far as their being bothersome in a physical sense, they were just about to get there when the drogue came out.

D9. Were you aware of lateral accelerations during reentry? If so, do you feel it is important to simulate these on the Centrifuge?

No!

D10. Was there any noticeable difference between the linear acceleration sensations experienced in the capsule and on the Centrifuge? If so, do you consider them important?

The linear g impulse during drogue and main is not a big acceleration that bothers you. The opening of the chute was a pretty good jolt but was a little mushier or softer than I thought it would probably be due to the stretchy nylon risers. Landing was a little bit sharper than I had figured it would be. I don't know whether you had this type sensation or not. I had thought that this nice big air ride bag back there as being a reasonably soft landing, but this was a pretty big jolt. It wasn't anything that is debilitating or anything like that, but it was just a little more than I had anticipated; a pretty good kick on the back, but nothing anywhere near severe enough to cause any difficulty. The biggest difference is that you don't have any tumbling on starting or stopping like you do on the Centrifuge where you have been going in rotary motion and when you stop or accelerate rapidly you have the feeling of going head-over-heels or heels-over-head, depending on which way you're going. There's none of this at all during reentry. The only time during the whole mission when I had any feeling of tumbling at all was just at SECO, at cutoff. I felt as though I had gone forward, head-over-heels, but this is different from the Centrifuge because your feelings of starting and stopping on the Centrifuge are very, very pronounced. You feel as though you do several loops before you really get the thing stopped.

E. Landing

E1. Was the capsule stable before drogue deployment? If not, describe motions.

We covered that in discussing same of our other material in Section D.

E2. Did you hear the drogue mortar?

Yes, there's no doubt when the mortar fires it's a very audible sound. It's not uncomfortably loud but it's a good solid sound.

E3. Did the drogue deploy automatically? If not, did you deploy the drogue manually? What failure indications did you have?

We described that adequately in Section D. The drogue did deploy automatically.

E4. Describe the drogue opening shock.

The opening shock was not a straight g pulse into the couch as we had thought it would be. This was due to the angle at which the capsule was yawed when the drogue came out. There was sort of a snapping effect as the drogue popped the capsule back up to an upright position. Did I deploy the drogue manually? I was just reaching for the button when it came out by itself.

E5. What was the indicated altitude at drogue opening?

We covered that in Section D.

E6. Did the drogue canopy "pulse"?

I didn't notice any pulsing of the drogue at all but there was a lot of back and forth motion of the drogue. I would estimate maybe 2 or 3 feet, it kept sort of ricocheting back and forth in a small circle.

E7. Did the capsule stabilize after drogue opening? If so, how soon?

Immediately. It was a matter of just 2 or 3 seconds until the capsule was upright and holding steady.

E8. Did the snorkel door eject properly. Did the cabin inlet and outflow function properly? At what altitude did each occur?

I don't remember hearing the snorkel door go at all, but I assume, that it did. That's one thing we can check on the capsule. I didn't get an exact altitude hack on the snorkel operation and this is another thing I'd like to check. I think if anything the snorkel probably operated a little bit low. I would guess maybe that it was down around 15 or 16 thousand.

E9. At landing did the vacuum relief valve function properly and did the snorkel valve prevent seepage?

As far as I know, they both operated properly and I didn't notice any water inside the capsule at all, not a bit. And I did check.

E10. Was the capsule stable when the antenna section jettisoned? If not, describe capsule motions prior to jettison of antenna section.

We described that somewhat in Section D. The capsule was very stable when the antenna section jettisoned and I could see the whole recovery system just lined up in one big line as they came out. It was almost just like its laid out in the pictures and the handbook.

Ell. Did you hear the antenna mortar?

Yes, and it sounded very similar to the drogue mortar *we* talked about a minute ago. These were not real loud sounds but they are very audible.

E12. What was the indicated altitude at antenna jettison?

I think we covered that in Section D. I believe it came out at about 10,800 feet where it's been operating on the pad.

E13. Did the antenna jettison automatically? If not, did you deploy the antenna section manually? What failure indications did you have?

Yes it did.

E14. Could you hear the opening of the main chute? If so, did the sounds coincide with the acceleration pulses?

No, I was looking right at the main chute when it opened and I don't recall any noise at all.

E15. Did any noticeable angular accelerations accompany the main chute opening?

No. I think the capsule was pretty well lined up straight below the chute; it was a straight g pulse into the couch but we had a pretty fair oscillation set, up after that. After main chute the oscillation wasn't large but we were swinging back and forth plus and minus 20 degrees at the maximum and the closer we got to the water the more this damped out. This is similar to some of the pictures we've seen of capsules coming down where they had more oscillation shortly after opening than it did close to the water. The chute was very stable by the time I got down to the water. I think we hit with almost no apparent swinging motion. At impact I could see no drift in looking out the periscope, I was watching through the scope and I could tell almost to the second when I was going to hit, because I could see the waves coming up very close right at the last. I could see every little whitecap and the detail on them, and just before landing I braced myself. So you can get a pretty good indication of height just by looking out the periscope. The action of the capsule at impact; it rolled to my right and down a little bit toward my feet down toward the hand controller or the number one battery. It went to that side momentarily and then collapsed over on its side like and then there was water all over everything. The capsule popped back up and assumed various angles from there on.

E16. Describe the opening shock to the reef condition.

This is a good solid g pulse, I don't know exactly what it was. I think we were estimating a maximum of 4g's and it probably was, a momentary 4g I would estimate.

E17. Describe your view of main chute deployment through the scope and the window. Did you see any chute or riser damage? Was the capsule turning relative to the chute? Estimate amount. Was the chute canopy stable?

The capsule was still turning after it was on drogue and it was still turning even after main chute came out. We were still rotating very slowly and this gradually stopped as we came down closer to the water and at impact we had no rotary motion at all. Did you see any chute or riser damage; no, I couldn't see any damage to the chute at all. All panels looked good, the risers all looked good, just like a completely normal chute.

E18. Describe the capsule motion after main chute deployment.

I think I've already done that.

E19. Did the main chute sequence light work normally?

Yes, it did.

E20. If it was necessary to use the reserve chute, what were the indications of main chute failure? What was the type of failure? Describe the reserve chute deployment. Could you hear the ejection bag inflate?

It was not necessary to use the reserve chute.

E21. Were angular accelerations noticeable at deployment?

I believe we've already covered that one.

E22. Was it a clean deployment?

Yes, very much so. It just looked like something out of the handbook all laid out in a big long row, and it blossomed right according to schedule. The chute unreeled according to schedule and filled out completely. I noticed no partial collapsing of any part of the chute it seemed to come out very evenly and came back in about 2 cycles and settled down.

E23. Did the impact skirt deploy normally and did the sequence light work properly?

I was going over my checklist and had not yet checked the landing bag at the time Cape CapCom called. I was getting the straps off as he called and said something about landing bag. I went to the switch as Cape CapCom started saying something about the landing bag; it was a reminder for me, and I went to the switch and by the time Cape CapCom had completed his statement I had flipped the switch and the bag as green. I came back and reported that the bag was green.

E23(a). Did you feel the bag was deployed at that time?

Yes, I could feel it release.

E23(b). Did you hear the squib fire?

I don’t recall hearing the squib fire. I can't say that it did or didn’t.

E23(c). Could you feel it?

Yes, you can feel it go down; there’s no doubt about that.

E23(d). Was there shock when it actually released or when it ended its travel?

I don't think I felt much when it went clear down. The main thing I felt was a bump at the time that you activate the switch. I don't recall feeling any bump as it hit the bottom of its travel at all.

E23(e). This was synchronized then with your movement on the switch?

That's right. What I felt was the initial part of the movement. I assumed at that time that it was the lugs on the heat shield releasing and this is just what it sounded like and felt like.

E24. At what time did the skirt deploy relative to main chute deployment? Did the heat shield drop have any shock effect?

Only at the start of the drop.

E25. Describe the voice communication with the ships and aircraft during the parachute descent.

They picked me up right away. I think the first estimate was that I would be in the water about 20 minutes, then they shortened that estimate, but I think it actually turned out to be about 20 minutes. I don't have an exact time on that. Communication was very good with the *Noa*, the Steelhead as it is called. I questioned whether they had me visually at the start and they said they did so I left the UHF VF in the RT position so I could continue to hear what they said rather than putting it back on normal for their homing. They had a P2V around me within a few minutes. Voice communication back to the Cape at that time was pretty spotty. I don't know whether the Cape CapCom was hearing me all the time but I wasn't reading him.

E26. What was the approximate capsule attitude at impact? Estimate impact acceleration. What was the approximate attitude and magnitude at second impact?

I think we covered most of that, the approximate attitude and magnitude at second impact. I don't think I can differentiate too much between the time the shield hit the water, when the capsule came fully down on it and submerged and then popped back up again as being a primary and secondary impact. This occurred pretty rapidly and you just feel it as one big g pulse. I don't really feel you can accurately say that you felt one impact here and you felt another impact there.

E27. On what part of the chute swing did landing impact occur?

I don't think I had much swing at that time. The direction that the capsule appeared to go to after impact may have been due to a real slight horizontal velocity, but if it was, it was not apparent to me in the periscope before impact. I don't know exactly what the winds were in the area; I would guess after I was out on the ship that they were probably 10 or 12 knots. So I probably did have a little horizontal velocity, but I could not pick it up in the periscope before impact.

E28. Describe briefly the landing impact (motions and accelerations). Compare with any other common experience, if possible.

I think we've already covered that one.

E29. How long did it take for the capsule to right itself?

The capsule popped right back out of the water and went over to the other side and pretty well stabilized at maybe 20 to 25 degrees off of vertical and almost directly to my left. It was going back and forth quite a bit in the waves but the general attitude it always came back to was a little bit to the left. Over a period of about some 10 minutes bobbing around; it gradually had straightened up quite a bit and was quite vertical, probably within an average position of maybe 5 degrees or so of vertical, and it was fairly stable around that attitude The question was what angle, how much angular travel did it make while it was bobbing around in the water, I think this probably was an average of plus and minus 20 degrees. Occasionally a big wave would hit it and it would go on over to maybe 50 or 60 degrees and bob back up again

E30. What was the final trimmed angle of the capsule in the water?

Answered in E29 above

E31. Did the main chute disconnect normally? Did the reserve chute eject normally? How could you tell?

As soon as I had bobbed back up again, the rescue aid light was red. I operated it manually and I didn't feel the main chute disconnect. I wasn't positive that the chute had disconnected for about 10 or 15 seconds, then I saw it through the scope in the water and I knew that it was disconnected at that time or at least it wasn't pulling the capsule anymore. I didn't have any sensation of the reserve chute blowing out at all or of firing, and I assume that it did. This package was lying in the water along side the capsule (I could see that through the periscope also), but I don't recall hearing any big firing of the reserve chute and it's surprising because these other pyros operating earlier were so audible that I would think I would have heard that one in particular. I operated this switch pretty rapidly after impact, though I think I caught it bobbing back up before we'd been on the water very long. There was quite a bit of noise from the water hitting the side and it may have blanketed some of this other sound from the reserve chute being popped out.

E32. Did any equipment break loose at impact?

No, there was nothing flying around inside at all during impact. The g loading at impact that's a difficult one to estimate because it's mainly a shock g and those are extremely hard to estimate. I would guess the initial shock is probably 12 or 14 g's but it's very short (probably not over 1/4 or 1/2 second duration).

E33. Could you anticipate the landing time and prepare for the landing shock?

Yes, I had gotten into position so that in case I was not estimating altitude properly through the scope and it sneaked up on me, I would be ready for it so I was back and pretty well braced. The closer I got to the water, the more detail I could pick out through the scope. I used the little hand mirror a couple of times but there was so much junk on the window that I couldn't get nearly as good a view of the water with the mirror as I did through the scope. I finally gave up trying to estimate my height with the mirror. One reason that I wanted the mirror was that I thought it would give us our best estimation of height above the water, but I just couldn't see enough through the window to make a good estimation. The closer I got to the water, though, the better the indications were through the periscope, and, just before impact, you could see the little white caps and every little wave detail and you can get braced very adequately.

E34. Could you estimate your horizontal speed at impact?

I think we've already covered that.

F. Post landing

F1. Comment on the period of time while you were waiting for recovery vessels or aircraft.

The ship kept telling me that he was 1,000 yards away and then closer in and closer in and I couldn't see him for a lot of this time. I tried to use the mirror again out the window and still couldn't see him, couldn't see him in the periscope, apparently he was coming around on the side away from the scope. Finally the capsule did turn enough so that I had the ship in sight through the scope and all the time he was making his final approach. The scope was lined up perfectly. If I had had control of it, I couldn't have lined up better with the ship so I just watched him coming down; it was a real good sight. This bobbing around out there as I mentioned yesterday was the only time I had any stomach awareness at all on the whole trip and this was after 10 to 15 minutes on the water. I wasn't sick or really nauseous but I was aware that I did not feel quite as good as I had felt before. I think the main comment on this period of time while waiting for recovery was heat. I was very warm. The capsule had heated up considerably inside. I was sweating very heavily at this time and was trying to remain just as still as I could so that I didn't add any more heat to this than was necessary.

F2. Comment on ECS cooling during the post impact period.

The humidity increase was very noticeable after being in the comparatively dry oxygen atmosphere all this time and then coming down close to the water. I was not cooling down as a result of the flow even though I did remain pretty static in the capsule during that period. There was no extra activity I had to do at that time. I did run through the postlanding checklists on the main chute and after impact. I went through each item first from memory and then by checking the checklist to make sure that I had everything. I had not completed this check when Al called about the landing bag on the main chute. On the water after impact I immediately began getting the straps and helmet off. I kept the neck dam pulled down over that little collar hook so we kept flow going out without blowing the suit up and that worked very well. I thought about leaving the helmet off to aid cooling a little but I didn't want to lose communications with the ship so I left it on.

F3. Was there any leakage into the capsule from any source? If so, did you do anything to stop it?

There was no leakage at all. I looked everywhere I could look to see if there was any leakage, especially after these chunks had been coming off during reentry. I had visions of these being possibly either a broken heat shield or something there that might puncture the capsule or burn through it a little bit. Actually, back in that area the only place you can look is in the area surrounding you in the upper part of the capsule where a leak probably would not occur.

F4. This question has been answered previously.

F5. After landing could. you tell the status of the following rescue aid: SOFAR bombs? Chaff? Beacons? Dye marker? Light?

No, I could not tell the status of SOFAR bombs. I knew chaff was out because the airplane said he had picked it up on radar; however, I did not see it. I did not know whether the beacons were operating or not. I could see the dye marker in the water while the ship was making his approach and before I had him in sight. I learned from the ship that the light was working on top of the capsule.

F6. How rapidly did the battery voltages deplete after landing? How did the current vary?

The battery voltages came down to about 25 volts shortly after the start of the flight and were still holding steady all the way through to impact. I didn't check the voltages just before leaving the capsule but I did turn the ammeter switch to off.

F7. Could you detect the erection of the HF antenna?

No, wasn't sure it was up until the ship said they had it and would cut it during their approach to the capsule.

F8. Was steam noticeable at any time?

No, I didn't see any steam at all.

F9. Did you think the heat shield was still hanging below the capsule while floating?

I assume that the heat shield was still down there or we would not have been floating in the attitude we were in.

F10. Did the explosive hatch eject properly? Also comment on the noise, etc.

I took the cap off and the pin out positioned my hand with my knuckles beside the plunger, turned my head back away and shielded my head with my arm and then bumped the plunger with my hand. That was a mistake because the plunger kicked back out and cut both of my knuckles through the gloves and threw my arm back over in the capsule. My knuckles were bleeding down both fingers when I finally undressed later on. I was afraid at first I might have broken them. I highly advise if anyone ever has to use that plunger again they do it with the meaty part of their hand or possibly take the swizzle stick and rap it from the side. The hatch made a good loud bang when it left but it wasn't uncomfortably loud. I still had the helmet on at the time so that undoubtedly attenuated a lot of the noise. I noticed after the hatch blew that quite a bit of gray paint from the inside of the hatch area flaked off and settled inside the capsule.

F11. Describe your egress from the capsule. Any problems or changes recommended?

Well, this is very much like climbing out of the Procedures Trainer or climbing out of the capsule on the pad. There were two men back behind helping lift me out and keeping me up above the edge of the hatch so we didn't rip the suit. Egress was completely normal; I pulled my legs around in the same position we have for getting in and out. I asked one of the men to hold his arm up so I could hang onto it. We've found climbing in and out of the capsule is a lot easier if you have a handle up above the hatch or someone's arm to hold. on to so you can hoist yourself and slide right in and out, and he did that.

F12. During egress did you encounter any hot spots on the capsule?

No, none at all although I still had my gloves and full suit on. Comment on this should probably come from the shipboard people.

F13. How much of the survival equipment was used? Was everything adequate?

I didn’t use any of the survival equipment. I did have the pack unstowed, however, and ready for use. I didn't have any trouble releasing the pack.

F14. Did you notice any deficiency in the status of training relative to capsule egress? Was the capsule more or less stable hydrodynamic ally than the Egress Trainer?

My status of training relative to capsule egress is good. I could have egressed from the top had it been necessary. Hydrodynamic stability is very similar for both. They have the same type motions and general banging around.

G. Recovery

G1. Were there any difficulties during the ship or helicopter pickup? Describe.

The ship drifted down on its first approach in very good shape. My initial physical contact between the capsule and the ship was a very gentle rubbing up against the side of the ship; I could tell when it happened, of course, but it was very gentle and he immediately said they had the shepherd's crook on it. There was very little delay in starting their hoisting operation. They pulled the capsule up part way out of the water and then let water drain out of the bag in that position. They pulled it up rather slowly for a little while, then hoisted it right on up and stopped in a position out of the water. I thought they did a fine, expeditious job of picking it up.

G2. Were any recovery ships or search aircraft sighted while the capsule was still descending on the parachute?

No. The only ship I saw was as I came over Area G. Looking down at the water there was a clear area in between some clouds and I saw a little "V" in the water. I couldn't see the ship, but the "V" in the water headed almost straight west. We can check what the ship's heading was at that time to make sure the ship was there. I went immediately back to my map to check and make sure that I was just about at the right time to be over the recovery area. I had just passed it and was looking back this way, down a little bit below orbit attitude in pitch, probably at a 50 or 55 degree angle; when I checked the time it was exactly the time when I should have been looking right back at Area G. By the time I had checked all this and looked back out the window again I couldn't see it again for clouds. So I think probably this answers our question on whether we can see ship wake up there or not; I think we could. I'd like to check and see what the ship's heading was at that time because if they were heading roughly west I think that was the wake of the ship.

G3. Was adequate information obtained from the recovery helicopter or ship to allow you to decide whether to egress or not before capsule pickup? If not, what was lacking?

Yes, information was adequate. I considered getting out but I decided that I would stay in, mainly because of the heat. Perhaps if I had started egress immediately after impact and had gone ahead and taken the increased heat right then by getting out rather than staying in the capsule I might have been better off. That's debatable; I don't know which was better. I elected to stay in and it worked out alright so I guess in this case, at least, it was OK.

G4. If you did egress prior to capsule retrieval, did you receive assistance from recovery forces and was it adequate?

We already covered that.

G5. If you remained in the capsule until it had been retrieved by recovery forces, did you receive proper and adequate assistance from shipboard personnel during capsule egress? Do you have any recommendations in this area?

I think it went just like we talked about and just like we briefed. I don't know anything that the ship could have done that would have been any better except just getting there in a bigger hurry. Their procedure was good, their approach was good, they hooked onto the capsule the first try and got. it out of the water in good shape. I think their recovery procedure was just like we had briefed on and it was very good.

G6. If the Stullken collar and the skin divers were used in your recovery, do you have any comments on their performance? Were there any communications problems?

The Stullken collar was not used; skin divers were not used, although the first report was that they were going to put the divers in the water, but they changed their mind.

G7. Narrative account of egress

When the destroyer came along side they hooked on with the shepherd's crook and someone told me he had cut the antenna. He was keeping me very well informed all during his approach. During the capsule pickup I got one real good bump, it was probably the most solid bump of the whole trip. I understand that a line slipped or they didn't have one of their stabilization lines pulled up. The ship rolled and swung the capsule out away from the ship and then rolled back and swung the capsule against the ship. It jammed me over to the side of the couch. But wasn't anything that would hurt me. After we were on deck there was some delay while they were doing something up on top; I never did find out for sure what they were doing. They asked if I could hear the men talking down from up on top. I could not; there was quite a bit. of inverter and other noises still going on in the capsule. I still was in good communications with the ship, however, on UHF. I took the panel loose I kept the suit hose connected and worked over and couldn't quite reach the last restraining strap for that hose extension that was stowed remember that way down by the end of the wire bundle on the right. I worked a little bit trying to get down to the extension hose and still keep the vent hose connected, didn't want to disconnect the vent hose until I had to. I probably added more heat to the system during that period straining against the vent hose restraining strap than if I had gone ahead and disconnected the hose and reached down and picked up the extension. But I finally pulled the extension loose without disconnecting the vent. I was working my way up to sitting up on the right side in the capsule, had my head up towards the back of the hatch and then disconnected the suit hose and was in the process of hooking up the hose extension when I decided to egress by the side hatch. I was really hot by this time. I was pouring sweat, and I decided that the best thing at that point to keep me in a little better shape was to come on out rather than come up through the top. If I had to get out through the top I could have, but I was extremely hot at that time and with the means of getting out I didn't see any point in fighting that one any further. I don't know whether you heard our communications with the ship or not at that time. I think he asked me whether I wanted to blow it or not. I said I did and to let me know when everybody was clear outside. He called me back and told me when everybody was clear.

H. Evaluation of Capsule Systems

H1. Comment on your suit. Do you suggest any changes?

Well, we'd all like to go in a shirtsleeve environment obviously; but the suit worked fine; I had no problems whatsoever. I think the only unusual thing with the suit was probably that I had not thought that the body posture in the suit would be particularly different under zero-g, but it is. When you're under zero-g there's a tendency to sort of go into a natural stoop. It may be that the helmet tie-down makes a sort of pull in that direction so it naturally bends you over a little bit if you're just relaxed in the capsule. I had to consciously put my head back in the couch when I wanted it to be back in the proper position. I know of no changes I would make in the suit right now.

H2. Comment on your couch. Changes?

If we had had more time to look into the changes on the couch before launch I think possibly I would have gone to the change I suggested back at the Cape earlier and that is just do away with the leg troughs. This would make a nice big area in which to put your legs, so you can move them around in any direction you wish or cross your feet, or put your legs over in one corner or the other. It would make more room for equipment. We want to run this through, I think, on the Centrifuge, or on the track at Holloman, or something like that. All we need, possibly, is just a restraining strap around the knee to keep it from going too far over into, the hand controller or over into the abort handle. I think this would let you move your legs around a lot more and still give adequate support, with just a net arrangement stretched across on top of the equipment in that area. This would give you not only more body motion area down in there and more area in which to put equipment, but I'm certain that it would make it easier to get the small pressure bulkhead down in there and stowed during egress. That is not the primary thing that holds up the small pressure bulkhead, though. The control handle and the area around the side of the periscope are the primary things that hold it up. I think it would probably still be easier to stow it down in that area. During the zero-g state, though, the leg troughs being in there cause no problem. You feel very comfortable. I don't recall ever having to move around in the suit to get rid of a little bunch of material someplace like you do sometimes when you're at one g and you've been in the couch for a while. It is just like you're floating in the middle of the suit. There just are no pressure points, a very comfortable feeling. The couch was adequate during the high-g period, which of course, is what it was designed for. I had no pressure points, or uncomfortable sensations from the couch at all during any of the accelerated periods. The restraint harness was adequate.

H3. Comment on your restraint harness. Changes?

We had removed the knee restraints because we felt we didn't need them. And I don't feel that I needed them at all on this, flight at any time during landing, or any of the motions or accelerations none of' them gave any vector that I felt would have been better with knee restraints. Shoulder straps were adequate. I could lock them okay. I locked them before coming back in again, before reentry, and I had them loose during the zero-g period and intentionally pulled them pretty well back up so that the spring was not pulling down tight on my shoulders.

I didn't release the lap strap; that one is too hard to get tightened up again properly. Even when Joe puts me in there, with him leaning and pulling on the straps very hard, we sometimes have a problem getting that tightened up to where I like it, so I didn't release that at all. I did have the shoulder straps good and loose, however, and the chest strap. This does help you move around some. It helps you move up toward the window. I don' t think the straps are the limiting factor in moving up to look out the window. I think it's mainly the suit.

H4. Comment on any difference in quality between HF, UHF, and command system outputs.

Well, there's no more difference in orbit communications between HF and UHF than you would anticipate between HF and UHF on the ground. HF is not as reliable, but it is longer range. It's spotty and more subject to interference. You hear some static on it occasionally that you don't hear on UHF. UHF is shorter range, but once you have UHF communication it seems you have it very solid with no problems. I either had it or didn't have it. It was usually pretty clear. I don't know how the HF checks were on the ground, of course, but when we made this worldwide HF check on the first orbit at Australia, I came up and made my call and I heard Cape CapCom right back. Cape CapCom was loud and clear and I heard all of the transmission, I think. It was real spotty elsewhere. I'd just hear a word maybe, once in a while from somebody else, but I never received any other solid transmission on HF during that whole pass until the Australian stations came up and I heard them. But for some reason, Cape CapCom was the only one I heard coming up on the net reliably. We switched back and forth, I don't know how many times I wouldn't even estimate how many times we switched back and forth between HF and UHF during the flight. After checking UHF low, we did just like we planned and left it off and flipped back-and-forth between UHF high and HF and turned the VOX switch off each time as we had planned, to make sure we didn't wreck the transmitter. That seemed to work very well. You get into a habit pattern on that, where you automatically go down to the transmit switch and automatically reach up and cut the VOX off before you do anything with the transmit switch. You almost tie the two together as far as operation goes.

* * *

John Glenn Jr (center) with T.J. O'Malley and Paul Donnelly in front of the *Friendship 7.*
credit: NASA.

With Guenther Wendt following a launch scrub. credit: NASA

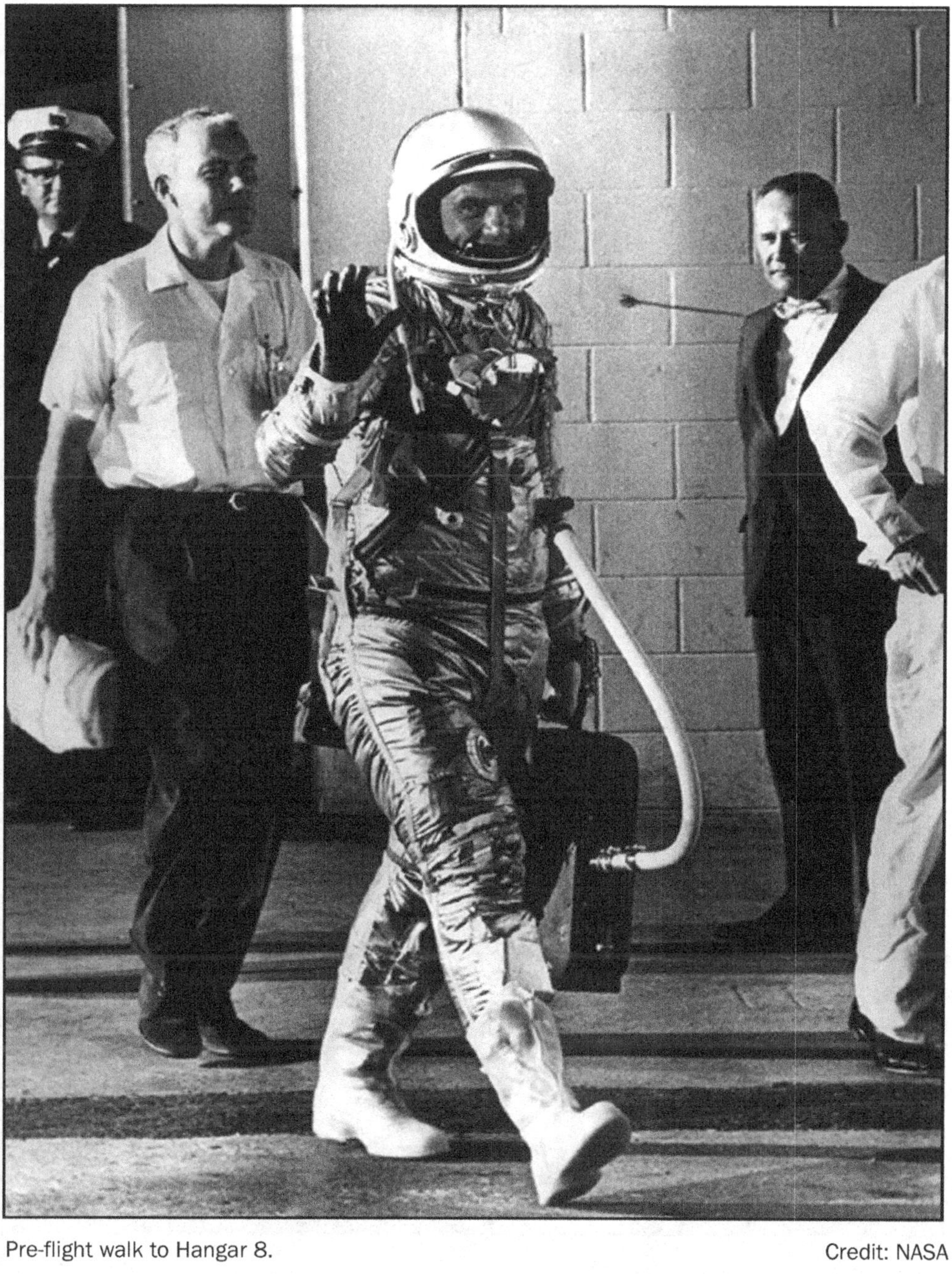

Pre-flight walk to Hangar 8. Credit: NASA

Friendship 7 brought onside the USS *Noa*. Credit: NASA

Relaxing on the USS *Noa*, 21 minutes after being brought onboard. credit: NASA.

Inspecting the spacecraft following flight, onboard the USS *Noa*. credit: NASA

Dictating notes into a tape recorder. Credit: NASA

Packing his spacesuit, onboard the USS *Noa*. Credit: NASA

Winched into a helicopter that would transport him from the USS *Noa* to the aircraft carrier USS *Randolph*.
credit: NASA

Being honored by President John F. Kennedy at the Manned Spacecraft Center. credit: NASA

Post-flight parade in Cocoa Beach, Florida. credit: NASA.

Official U.S. Senate portrait.

36 years after he became the first American to orbit the Earth, John Glenn jr became the oldest person to fly in space by serving as a payload specialist on STS-95, on the Space Shuttle *Discovery*.
credit: NASA

Made in the USA
Monee, IL
04 November 2026

67387898R00049